First edition: April 3, 2023

Copyright © *2022 Marcos Cervantes Janssen*

Edited by Editorial letter@red

https://www.youtube.com/channel/UCQ12Xlt8oQOaWAhAiboXPUA

https://www.instagram.com/newtekjanssen/

https://www.facebook.com/LETRA3ROJA

https://www.newtek.janssen@gmail.com

https://twitter.com/Letra3Roja

https://newtekjanssen.es.tl/

letra3roja@gmail.com

PATENT
ISBN: 9798390351673

WHEN, WHERE AND HOW.

By: Marcos Cervantes Janssen

INDEX:

- FOREWORD: 5
- INVENTOR: 7
- INNOVATIVE: 9
- ENTREPRENEUR: 11
- AUTHOR: 13
- CREATOR: 15
- DESIGNER: 17
- COMPOSER: 19
- INVESTIGATOR: 21
- DRIVING: 23
- CONSOLIDATOR: 25
- GIVER: 27
- MENTOR: 29
- VISIONARY: 31
- ISBN 33

FOREWORD:

This work is the practical sample of a patent, because in this book I will show you that you can patent your idea, project or device through Desktop Publishing; I will give you all the necessary tools so that you can achieve this task. Patenting has been throughout history, the way in which our technology evolves legally and fairly for creators. It is vitally important to immediately patent any original and innovative idea that you have, since the passage of time has left great inventions forgotten, and so have their original authors. Read carefully and clearly each of the recommendations, only by practicing and doing it, you will be able to complete this task to transcend, contributing your intellect and ability.

We are a single interconnected system and truly all of us without exception need others. Ideas arise in moments, as well as fleetingly vanish, which is why documenting transcends the human mind through time, today may be the precise moment to transcend, leaving a useful legacy, documenting in a concise and valid way. Each mind that arrives on the planet is capable of contributing to the collective evolution of our civilization. Each of Yours ideas is potentially transformative, present to the world, this personal or collective contribution, in a truly practical way. I thank you for your attention and I am sure your idea will be of great benefit to all of us, without further ado for the moment we will go to the first chapter, remember if you want to patent, this is the precise place, the moment and the most appropriate personal way to do what ALREADY!!!!!!!!

INVENTOR:

When the unknown becomes known, the invention arises, if you have ideas emerging from your mind, and you are excited to express them to others, your path is that of an INVENTOR. The solution to thousands of daily and specific problems requires inventors determined to break the silence and shyness. When suddenly in your mind you visualize the hypothetical solution, and experience the possibilities of functional success, the inventiveness in you is active. Now, if you carry out the experimentation of this invention shaped in your mind and it is successful, the inventor needs to patent such invention for its conservation and legitimacy. The word invent is the entrance to an unknown window, and thus go through the adventure of mental exploration, with the clear purpose of solving and interacting with a need, problem or desire to discover. Inventing has been in history the daily action of progress. Having new technologies, theories and hypotheses arise from constant human inventiveness. You are an Inventor, but you do not take it for granted, because it seems like a path that is too difficult and complicated, in this book, which is a Patent in itself, you will see the possibility of exercising the same opportunity that is available to each one of us; that is, to be a legitimate and true inventor. This way you will have a real opportunity to transcend, by reason of the invention at the service of others, remember that each problem has more than one solution, wanting to be discovered by an inventor of courage and determination. Entering this window of the future is the task of those who have their eyes on the future and their hands on the present, without forgetting the past as an inventive experience.

INNOVATIVE:

When an idea for improvement arises, as a result of a solution already given, it is called as innovation, each new edition of a book is an innovation of the original title, this in the industry is called as revision in innovative models for better performance . Innovation is not an invention that arose from nothing, but it is equally important, since continuous improvement in all aspects leads us to efficiency through excellence. The innovation process demands a high degree of analysis and proposal for improvement. Innovation is the essence of technical evolution in the industry, as well as continuous improvement in almost all administrative and technical processes. To innovate is to get involved at the design level, to obtain a new revision, version or edition, as the case may be.

To innovate, we must invite the most possible alternatives, in the practical and real solution, of the problem in question, this is how efficiency and practicality are elevated in their warmth and operation, through the innovation processes generated. The patenting technique, through bibliographical edition, allows the innovation of your patent, through new editions of the patented title. each revision an improved, expanded, and revised edition. Innovating the way to expose our patent is the essence of this work, which when understood and assimilated, will undoubtedly lead us to the solid construction of our foundations and first steps in this new era of transcendental challenges. He clarified that the primary invention does not derive but from the direct mind of the Author and inventor in question.

ENTREPRENEUR:

A true entrepreneur is not afraid of failure, because as an entrepreneur he understands that being an inventor or innovator needs courage and encouragement to navigate this adventure. To undertake is to begin, so as we know; each beginning requires an effort of great demand and challenge. Discovering what new horizons have been forgotten is not pleasant, which is why entrepreneurship requires great perseverance and cunning. Starting always requires extra energy, so entrepreneurship, like a starter, begins with a creative big bang, this being fundamental in evolution and dynamic change; For this reason, undertaking is a creative and willful action with futuristic purposes, through an orderly, energetic, and purposeful present.

Being an entrepreneur is essential to create a new patent, as well as editing the literary work for its description, and registration. In this way you will get a **ISBN, (International Standard Book Number),** with which the idea already has an author and intellectual property. Literary entrepreneurship is the most practical and viable way to establish authorship of your ideas, designs and inventions. This work which you have in your hand represents, in essence, such a function. This book is the undertaking by which patents can be realized in a practical and direct way. Every undertaking that you carry out,document it immediately in drafts in the first instance, with the clear objective of making them public, under his authorship. So in this way you will be the Author and holder of the legally registered rights of what is written and documented.

AUTHOR:

By ordering your ideas in a personal and automatic way, according to your previously practiced strategy, you will be able to be the sole, exclusive author of such ideas in a real and prompt way. Being an author is truly important for your personal development as an inventor. Thus, by practicing the authorship of your projects through the correct literary documentation, it will lead you to own your patent. This is so, internationally, thanks to the attributes of the **ISBN**, **(International Standard Book Number).** It is thus in this way that in this 33-page work he will take you to experience the authorship of his projects. Being builders of our civilization is an exceptional task; this in any of the existing areas and to be created. Remember that being an Author is universal in nature.

Creative and inventive authorship involves all fields of research and technological development, as well as the artistic area. Within all these branches to be developed, there will be multiple topics to document, such as; science, music, poems, medicine, psychology, specialties, etc. Being an author is the one who promotes his own disclosure in order to share his being. Only as a true author can you give yourself to others. The acquired and experiential knowledge that resides in you, you can by your own means, transfer to others, this being one of the maxims as human beings. Each solved problem is worthy of capturing generational significance, and it is through writing that it lasts the longest without deviations and dilution. Being an author also entails learning forever as a lifestyle, it is in this way and only in this way that the mastery of life becomes ours.

CREATOR:

A creation consists of the interactive correlation of elements and ideas, components and composition, it is in this way that matter takes shape, that scores are transformed into melody. To be a creator is to comfort individuality, an organized and functional system, it is to unify through intelligent couplings, a structure with its own identity. Creating is not an act of spontaneous generation, but rather a profound evolution of materialized ideas. Being a creator is procreating for others, contributing to the common good, giving birth to solutions and expressions that transcend time and space; Thus we are creation and creationists in this eternally evolving existence.

CREATING IS BELIEVING AND GIVING TO OTHERS WHAT arise FROM OUR INTERIOR.

The property of Creator is awarded to the sovereign entity, which humanity has identified as GOD, with this we can see how important humanity considers the exercise of such activity, with which it seems that creating is of divine nature, as well as us as humans we are deities which purpose and virtue we possess by inheritance. Creating is the only way to evolve in an integral way, thus taking our civilization beyond the already known earthly desires. Patenting each of our creations is our right and an obligation, this in favor of the order that caused the new human civilization; which is created, creative and renewing, in its eternal walk. Thus, we are humans with an awareness awake to progress, in this analogous existence, full of challenges to solve.

CREATING is BELIEVING in truly POWER.

DESIGNER:

Designing is the task of directing our creative thoughts in an orderly manner. The designer directs those dreams, which have not yet been realized, through strategies, methods and tools for the conformation and consummation of this. Being a book designer requires work, more than virtue. When your patent is embodied on sheets of paper, the design of this material is essential for the clear representation of the patent in question. The design ranges from the birth of the idea to the consummation of the patent given in question.

Documenting has always been the most substantial way to inherit wisdom, in this case the design is paramount because the accuracy and precision determine the editorial coverage of the patent. Each documented design is highly reproducible, so much so that it is possible to commercialize ideas through this method of patenting. Designing a book is the best way to practice editorial patenting. A Designer is a planner par excellence, his designs being premeditated creations with a high degree of awareness and active vision.

THREE WRITTEN WORDS SAY MORE THAN A THOUSAND WORDS ON AIR.

COMPOSER:

Within the subject of patenting, music has a very important place, since it is through it that culture and education are transmitted from generation to generation, writing music is a technique, which requires special knowledge. It takes attention and a lot of time to master sheet music writing, thus putting music on paper for hereditary preservation.

Composing music entails the art of expressing mental situations with melody and the technical, grammatical ability, this feeling as a precise idea. As well as writing, it is also very important to read and interpret lines of music as faithfully as possible. This is how the patenting of melodies is carried out only through written scores. The transmission of melodies through sound or merely manual training, loses its accuracy from one generation to another, not so when it is embodied on paper, it is possible to reproduce and preserve the complete authorship of the symphony in its entirety. Interpreting music is a sublime art, but composing music is an incomparable vocation and virtue for human development and its history, this is how each culture, region and social group imprints its feelings on this path of existence.

COMPOSING IS CO CREATING THE BEAUTY OF EXISTENCE.

INVESTIGATOR:

Research methods are the pattern of technological discovery of our history as humanity, research is in us, pure nature for our perpetuity. It is for such a reason that our mind always will search, reinvent everything you learn. Every time we analyze information, the search in the verification and breadth of such a topic is turned on in us. Corroborating is our task as natural researchers, verifying each situation and information, reaffirms our knowledge of things and situations.

IT IS NOT THE ONE WHO KNOWS MORE, BUT THE ONE WHO DOES MORE, THAT TRANSFORMS OUR SOCIETY.

Researcher is the one who does not stop and just observe, but the one who dynamically collects information and, under an order and strategy, manages to reveal results. The research methods, as well as the personal skills to develop them, coexist under the same vision, of creating today in the present, based on the analyzed past, a future structured under the guidelines of purification and improvement. Researching in a natural way is correct, plus planned research, based on experimental structures, will undoubtedly give better results and in less time. Only based on permanence and dedication, *a patent is always the successful culmination of painstaking and painstaking research*.

DRIVING:

The impulse received to undertake an innovation comes from different sources, the most important and permanent being one's own personal interior. Whether by longing or external influence, each impulse in us should in turn be transferred to others. Thus, as promoters of our own reality, we will be able to foster the spirit of others. A being who is a true promoter, promotes and infects his peers, to walk renewed again. This work drives your decision to take your inventive intellectuality seriously, and through documentaries in writing like this, you generate innovations, ideas and improvements under the scheme of internationally registered books. I invite you, through writing, layout, and editing your project, generate literary works that are read worldwide, under the different copyright protections offered by the To edit, today as independent authors. Remember each finished project, will lead with experience, to the resolution in its opportune moment. The experience of using the methods and equipment designated for this purpose, give patent results of impulse.

WE PROMOTE THE CREATION OF PATENTS THROUGH THE EDITING OF DIGITAL BOOKS. (ISBN)

CONSOLIDATOR:

If you've gotten as far as this line of reading, it's time to cement your idea as your own patent. Do not hesitate any longer and carefully write your invention, discovery, innovation or creation, whatever the genre, so in this way, consolidate on paper what has been around in your mind for a long time.

Remember that hypotheses are also subject to patenting.

We know that the scientific method requires experimentation for its validity, but it is not essential for its consolidation as a patent subject.

That is why I recommend that you immediately make your own and safeguard your intellectual property before time, plus daily occupations, distract you. Mental dissipation causes the consolidator not to complete his task, for this reason it is highly recommended to shorten the decisive times, and react immediately, to the expression of ideas on paper, do not leave at stake what can transcend and change what is necessary for the common good. Remember that closing the cycles is always of vital importance to continue evolving and transcending through the responsible consolidation of our actions, decisions and creations. **With solidity, success is more certain.**

GIVER:

It is the main theme of the creators, since the making of contributions is based on giving oneself and then receiving what corresponds by cause and effect. If your contribution to patenting is of this nature, it is vitally important to offer it in writing and reproduced for the knowledge of the largest number of beneficiaries. Remember that a patent is a giving, to others, with the clear reality of royalties; more so that the motor of the invention is the continuous improvement of our species. In this way it is as a result, the fair profit for our well-being and that of our community.

Giving the best of our intellect in self-publishing will, by its very nature, promote the patent in question, as well as the indications for its use and reproduction. This book is specifically designed to instruct through the editorial patent, the reality of it. It is checking that this idea is true and practical, it is the DAR of the work, for each of the potential creators. Thus, the opportunity to patent is given in a practical, simple, prompt and real way. Giving your time to learn how to write and design the correct design for each of the patents is the individual task of each person with this wonderful profile.

Remember this human truth, **GIVING IS LOVE.**

MENTOR:

If the idea to be patented is already clearly in your mind, make a draft, step by step, with all the details, because you will be the mentor of whoever has to understand your patent at the time. Each patent is a legacy for others, so the teaching service is exercised by vocation. Always remember that as part of a whole, you and I learned, which is why it is our duty to teach, guide and educate the new generations. Thus, as mentors, we can make use of this very important tool that is the publication of literary works. Every mentor requires a reliable source as backup, so a book is indicated.

The mentor is a teacher who guides and ensures that learning lands in a real practice, since he accompanies the disciple in question for as long as necessary. A mentor also offers his face as a friend and not only as a teaching authority. A mentor becomes a family through this royal path of learning, for this reason their relationship goes beyond objective information. Each transferred experience must thus be accompanied by prior experience, which is why each writing becomes a mentor when it becomes legally valid and recognized for its influential existence. A patent is a mentor for those who make it to innovate their current knowledge.

VISIONARY:

One of the biggest motivations that you have surely experienced is to reveal what you have innovated or discovered inside, to help or demonstrate that this or that situation has a solution. Being a visionary is seeing far beyond our own interests, which of course are a priority in everyone's life. Every time you face a daily problem and find a new solution in your mind, your brain will be projecting a better future in your thinking, and this not only for the individual good but also for a broad and extensive common good.

Planning for the future and being a strategist is truly being a visionary clinging to real and tangible results. The vision is one that is formed in our mind, but it is real, in the way that we carry out acts for its corroboration, and practical use. Being a visionary on the subject of patenting is leaving a legacy available to everyone. A discovery, design or conformation requires a complete vision of our lives. This is achieved by learning from the past, correcting in the present, and always living a future created by ourselves. You can only balance the present with its two components on its sides, past and future, experience and planning, foundation and ceiling.

ISBN:

As an epilogue I will talk directly about the ISBN, **(International Standard Book Number),** Which in this same book that you are finishing reading, is printed on its back cover, in this case this number is **ISBN: 9798390351673** and it is already part of the web, www takes it to the whole world and at all times, thus disseminating it is an inclusive task for our primary objective. We will be able to patent the largest number of innovations, discoveries and technologies, through worldwide and permanent registered edition treaties for independent cars. Low cost and universal coverage. An ISBN is single and personal, is the ID of a work in its entirety, making the author the owner of said document.

Remember that in a treatise, essay or writing, may capture in detail, everything that corresponds to your idea, device, innovation or hypotheses, data diagrams and the rest information, under one number Registered worldwide and valued worldwide as the intellectual property of the author related to said ISBN. Below I leave the QR for registration, payment and thus obtain it.

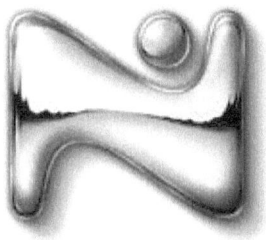

All rights reserved. Under the established sanctions

in the legal system, it is strictly prohibited,

without the written permission of the owners of the *Copyright*©

the total or partial reproduction of this work by

any means or procedure

reprography and treatment

computer.

www.ingramcontent.com/pod-product-compliance
Lightning Source LLC
Chambersburg PA
CBHW031558210526
45464CB00003B/1338